SOTHIS PRESS

Sothis Press aims to completely reimagine the public domain book with an emphasis on universal access to captivatingly designed print editions. We believe that technology will enable every single book ever published to be perpetually accessible in beautiful print form. Sothis is beauty in ubiquity.

www.sothis.press
1 Waterhouse Square, London, EC1N 2ST

Real and Complex Number Systems

Thesis

Elizabeth Ruth Bennett

SOTHIS PRESS

ISBN 978-1-0365-1510-2

This edition is published by
Sothis Press © 2023. The company
is registered in England and Wales.
The Company Number is 14900389.

REAL AND COMPLEX NUMBER SYSTEMS

BY

ELIZABETH RUTH BENNETT
(A. B., OHIO UNIVERSITY, 1903)

THESIS

FOR THE

DEGREE OF MASTER OF ARTS

IN

MATHEMATICS

IN THE

GRADUATE SCHOOL

OF THE

UNIVERSITY OF ILLINOIS

1908

UNIVERSITY OF ILLINOIS

May 28 1908

THIS IS TO CERTIFY THAT THE THESIS PREPARED UNDER MY SUPERVISION BY

Miss Elizabeth Ruth Bennett

ENTITLED Real and complex number systems

IS APPROVED BY ME AS FULFILLING THIS PART OF THE REQUIREMENTS FOR THE

DEGREE OF Master of Arts in Mathematics

G. A. Miller
Instructor in Charge.

APPROVED: E. J. Townsend

HEAD OF DEPARTMENT OF Mathematics

Table of Contents

Chapter I

Chapter II

Chapter III

Chapter I

If r be any positive integer greater than one, a positive number N may be expressed in the form $N = ar^n + br^{n-1} + \text{-----} pr^2 + qr + s$ in which the coefficients a, b, c, -----, s are positive integers and each less than r. From this theorem we know that we may select any positive integer as a radix or base and to each base will correspond a number system. Theoretically, then, it would be possible to have an indefinitely large number of different systems in which positive integers are used as bases. We shall attempt to consider only the ones which have been in actual use and those whose adoption has been advocated by mathematicians.

The first number systems developed slowly, forming gradually as the number concept of primitive man grew. In general, what are known as the natural systems, in which five, ten, or twenty are used as bases, seem to have developed most extensively.

The quinary is one of the simplest and most primitive in its structure of the natural systems. It developed among the races where only the fingers of one hand were used in counting; the number five being reached, the savage would use some expression meaning five and one more, thus unconsciously forming a system to the base five. The Betoya language of South America gives us one of the purest examples of a quinary system. Generally, however, the quinary is not distinct from the decimal and vigesimal. In fact, it is very rarely found as a pure system.

Conant in his "Number Concept" has given a list of the different tribes having a pure or mixed quinary system and a comparison of these lists shows that it is quite extensively used. Every part of the world except Europe furnishes us examples of peoples who have used or are using such a system. No quinary system has been found in any living European tongue nor does any earlier language show a trace of it except, perhaps, the Greek in the word πεμπάζειν, meaning to count by fives. The Roman notation, it is true, suggests quinary counting in IV, V, VI, etc. but the Latin language does not contain anything which would prove conclusively this tendency.

Ordinarily, the quinary system will develop into the decimal or vigesimal and we will have a mixed system. In Africa, Oceanica, and some parts of North America, the quinary almost always unites with the decimal while in other parts of the world, the quinary and vigesimal seem to have an affinity for each other.

When considering the principal vigesimal systems of the world, we note that this system was universal among Celtic races. The Breton, Irish, Welsh, Manx, and Gaelic contain well-defined vigesimal scales. It is also interesting to note in this connection that the French language contains a persistent vigesimal element as shown in the words soixante-dix, quatre-vingt, etc. With the exception of the word soixante, the French system is wholly vigesimal from sixty one to ninety nine inclusive. Other examples of the vigesimal method of counting are found among the Basques, the Danes, a few tribes in Africa, and a larger number of tribes in Asia, but by far the greatest number of tribes using this system are found in North and South America.

The decimal system has an origin similar to that of the quinary and vigesimal, beginning with those peoples who used the fingers of both hands for counting. It has become the most widely used of any single system.

Besides the natural bases which we have considered briefly, various other numbers have been used as bases or their use advocated. Leibnitz proposed a binary system of numeration in which the symbols used would be 0 and 1. The number 2 would then be symbolized by 10; 3, 4, 5, 6, 7, 8, etc. would be symbolized in this system by 11, 100, 101, 110, 111, 1000, etc.

The binary system, as advocated by Leibnitz, did not receive much attention, but instances are known where such a system has been in use by primitive peoples, as by certain tribes in Australia, South America, and Tasmania. Its origin among these tribes is thought to be due to the habit of counting in pairs.

An occasional ternary trace is found in some number systems. Such ternary traces have been found in the number system of the Haida Indians of British Columbia and in the systems of certain tribes in India and in Australia, but no pure ternary system is known.

Aristotle advocated a quaternary system and there are different instances of such a system being in actual use. Quaternary traces are found among the Indian languages of British Columbia, among the languages of certain tribes in South America and among the Hawaiian and some other languages of the Pacific islands. The Hawaiian language furnishes a complete example of the quaternary-decimal system, the quaternary element having modified the entire system which was originally decimal.

There seems to be no recorded instance of either 6, 7, 8, or 9 being used as the base of an entire number system, but there are instances in which a few numbers of a system to an entirely different base show traces of counting by either 6, 8 or 9. However, eight has been advocated as a suitable number base, it being claimed that the octonary system would be a much simpler one than the decimal.

Twelve has also been suggested by mathematicians as a base which combines within itself many advantages. It is said that Charles XII of Sweden advocated its establishment and in the seventeenth century we find Simon Steven of Bruges advocating the use of the same system. Buffon also remarks that twelve may very well be used as a base on account of the number of its divisors giving integral quotients. A duodenary system is in actual use in measurements of length and of quantity. For length measurements, we have: 12 in. make one foot.

For measurements of quantity, we have:

12 units	=	1 dozen
12 doz.	=	1 gross
12 gross	=	1 great gross

The old Roman metrology also used twelve as the base and in ancient French measurements, we find the following table:

1 foot	=	12 inches
1 inch	=	12 lines
1 line	=	12 points

One of the largest numbers used as a base appears in the sexagesimal system used in measuring time and angular magnitude.

Cauchy in considering how the present number system could be simplified without changing the base proposed to reduce the number of symbols by giving to each symbol two significations; one additive, the other subtractive. 1, 2, 3, 4, 5, 6, 7, 8, 9, 10, 11, 12, etc.

in such a system would be 1, 2, 3, 4, 5, $1\bar{4}$, $1\bar{3}$, $1\bar{2}$, $1\bar{1}$, 10, 11, 12, etc. The multiplication table would be much simplified and would be as follows:-

1	2	3	4	5
2	4	$1\bar{4}$	$1\bar{2}$	10
3	$1\bar{4}$	$1\bar{1}$	12	15
4	$1\bar{2}$	12	$2\bar{4}$	20
5	10	15	20	25

This system would introduce negative number concepts at the very beginning and, as we have stated, would simplify multiplication, but it seems questionable if there would be any great gain from its use.

L. Lalanne has applied the same method of notation to a ternary system. Unity in this system is represented by the symbol ϙ. The numbers of the system then are 0, ϙ, ϙϙ̅, ϙ0, ϙϙ, etc.

Besides the systems to different bases which have been enumerated, it has been suggested that a number system could be constructed in which no stated base would be used, the numbers being made to progress in their natural order or succession. In other words, every number may be made to serve as a base for a certain period of time. The originator of this system, Mr. Eissfeldt, does not claim for it any practical value but presents it simply as a mathematical curiosity. The system is formed by corrections, each number being used as a base on which the correction is made. A few examples will show the principle on which the system is constructed.

1	corrected by	0	=	1
1	" "	1	=	2
2	" "	0	=	3

2	corrected by	1	=	4
2	" "	2	=	5
3	" "	0	=	6
3	" "	1	=	7
3	" "	2	=	8
3	" "	3	=	9
4	" "	0	=	10
4	" "	1	=	11
4	" "	2	=	12
4	" "	3	=	13
4	" "	4	=	14
5	" "	0	=	15

etc.

The first twenty letters are then taken to represent the first twenty numbers and 0 is represented by X. Then we have:

ax = 1	cb = 8	fc = 24
aa = 2	cc = 9	fd = 25
bx = 3	dx = 10	fe = 26
ba = 4	-------	ff = 27
bb = 5	fx = 21	-------
cx = 6	fa = 22	230 = tt
ca = 7	fb = 23	231 = fxx

We shall now attempt to compare with the decimal system, systems constructed to the bases which have been advocated by different mathematicians, namely, the binary, octimal, and duodecimal.

In favor of the binary system, we may say that the only fundamental operations of any extent needed in this system are numeration and notation. Only two symbols are used, 0 and 1. Both these characteristics lead to considerable simplification, but the system is apt to become cumbersome on account of the number of places required for expressing a number in it, approximately about three and one-third times as many as in the decimal.

Eight, as a number base, has the advantage that successive binary divisions may occur without producing fractions while ten admits of only one binary division without fractions occurring. Its multiplication table is also much simpler than the decimal. Two symbols less are required and only approximately ten-ninths as many places are used in expressing a number in this system as in the decimal.

In 1859 Alfred B. Taylor of Philadelphia gave a somewhat lengthy report on this system before the American Pharmaceutical Association. The object of the report was to discover some system which would be more practical than the decimal for measurement purposes. The octimal system in its more elementary operations has been quite fully worked out by George Cooper and presented in "Octimal Notation," one of the books of the Western Mathematical Series.

The base of the duodecimal system, as has been mentioned, has the advantage of having the largest number of divisors giving integral quotients of any number that has been suggested by mathematicians as a base. For this reason ordinary fractions with denominators three, six, nine, or twelve give terminating radix-fractions in this system while these same common fractions do not give terminating decimals. Two more symbols are required in this system than in the decimal and the multiplication table is more complex.

The method of expressing any positive integer N to any integral base r greater than one has been stated in the previous chapter. In this chapter it will be proved that any proper fraction $\frac{A}{B}$ in lowest terms can be expressed in the form $\frac{A}{B} = \frac{p_1}{r_1} + \frac{p_2}{r_1 r_2} + \frac{p_3}{r_1 r_2 r_3} + \text{-----} \frac{p_n}{r_1 r_2 \text{---} r_n} + F$, certain restrictions being placed on p_1, p_2, ----, p_n. Several theorems regarding such fractions will be given. It will also be shown that radix and decimal fractions are only special cases of fractions expressible in the form given above and certain theorems regarding decimals will be proved.

Any proper fraction $\frac{A}{B}$ in lowest terms can be expressed in the form $\frac{A}{B} = \frac{p_1}{r_1} + \frac{p_2}{r_1 r_2} + \frac{p_3}{r_1 r_2 r_3} + \text{---} \frac{p_n}{r_1 r_2 \text{---} r_n} + F$, where $p_1 < r_1$, $p_2 < r_2$ -------- $p_n < r_n$ and $F = 0$ or can be made as small as we please by taking a sufficient number of the integers r_1, r_2, r_3 ----

$$\frac{A}{B} = \frac{Ar_1}{Br_1} = \frac{\frac{A}{B} r_1}{r_1} = \frac{p_1 + \frac{q_1}{B}}{r_1} = \frac{p_1}{r_1} + \frac{1}{r_1} \cdot \frac{q_1}{B}. \quad \text{But } \frac{q_1}{B} = \frac{r_2 q_1}{r_2 B} =$$

$$\frac{\frac{r_2 q_1}{B}}{r_2} \qquad \frac{p_2 + \frac{q_2}{B}}{r_2} \qquad \frac{p_2}{r_2} + \frac{1}{r_2} \cdot \frac{q_2}{B}.$$

Finally, $\frac{q_{n-1}}{B} = \frac{p_n}{r_n} + \frac{1}{r_n} \cdot \frac{q_n}{B}$

Substituting, $\frac{A}{B} = \frac{p_1}{r_1} + \frac{p_2}{r_1 r_2} + \frac{p_3}{r_1 r_2 r_3} + \text{--} \frac{q_n}{r_1 \text{---} r_n B}$

When r_1 r_2,----,r_n are given, the resolution of $\frac{A}{B}$ can be effected in only one way.

Assume the resolution can be effected in two ways.

Then $\frac{A}{B} = \frac{p_1}{r_1} + \frac{p_2}{r_1 r_2} + \text{-------} \frac{q_n}{r_1 \text{---} r_n B}$

and $\frac{A}{B} = \frac{p_1'}{r_1} + \frac{p_2'}{r_1 r_2} + \text{-----} \frac{q_n'}{r_1, \text{---} r_n B}$

Therefore, $\frac{p_1}{r_1} + \frac{p_2}{r_1 r_2} + ---- \frac{q_n}{r_1, ---, r_n B} = \frac{p_1'}{r_1} + \frac{p_2'}{r_1 r_2} + ---- \frac{q_n'}{r_1 ---r_n B}$

Multiplying by r_1, $p_1 + \frac{p_2}{r_2} + --- \frac{q_n}{r_2 --r_n B} = p_1' + \frac{p_2'}{r_2} + --- \frac{q_n'}{r_2 --r_n B}$

Therefore, $p_1 = p_1'$ since the integral parts must be equal. The same process can be repeated to show that $p_2 = p_2'$, etc.

The necessary and sufficient condition that q_n vanishes and the series $\frac{A}{B} = \frac{p_1}{r_1} + \frac{p_2}{r_1 r_2} + \frac{p_3}{r_1 r_2 r_3} + ------ \frac{q_n}{r_1 ---r_n B}$ terminates is that $(r_1, ----, r_n)$ be a multiple of B.

Let $\frac{A}{B} = \frac{p_1}{r_1} + \frac{p_2}{r_1 r_2} + ------- \frac{q_{n-1}}{r_1 ---r_n}$

The right hand member of the equation may be made an integer by multiplying by r_1, ----, r_n. Then the left hand member must also be integral when multiplied by r_1, ----, r_n. However, the only condition by which the left hand member can be made integral is that r_1, ----, r_n be a multiple of B. It is very evident that this condition is sufficient.

Since the (n+1)-term vanishes when $q_n = 0$, the number of terms in the series representing $\frac{A}{B}$ is the same as the number of factors in r_1 r_2, -----, r_n.

If B contains factors of r_1, ----, r_n, as well as other factors, the common factors may be removed and after a certain number of terms, the fraction will have the same series as some fraction whose denominator is prime to B.

Let $\frac{A}{B} = \frac{A}{r_1 r_2 B'} = \frac{p_1}{r_1} + \frac{p_2}{r_1 r_2} + ----- \frac{q_n}{r_1, ---, r_n B'}$

Multiplying both members of the equality by r_1 r_2, we have,

Integer $+ \frac{A}{B} = p_1 r_2 + p_2 + \frac{p_3}{r_3} + ----- \frac{q_n}{r_3 --r_n B'}$

Then $\frac{A}{B'} = \frac{p_3}{r_3} + \frac{p_4}{r_3 r_4} + ---- \frac{q_n}{r_3 r_4 ----B'}$

The same method of proof applies if B contains any other factors of $(r_1, ---, r_n)$.

Any rational fraction $\frac{A}{B}$ in lowest terms whose denominator is prime to $r_1, ----, r_n$ is periodic.

Since the series representing $\frac{A}{B}$ is obtained by always using B as the divisor the process must recur, there being only B remainders possible.

The fraction $\frac{1}{B}$ will now be considered and its period will be determined in respect to the series $r_1, ---, r_n$. It can then be shown that the length of this period is independent of the value of the numerator.

Finding the period of $\frac{1}{B}$ with regard to $r_1, ----, r_n$ consists in finding the exponent to which $r_1, ----, r_n$ belongs, mod. B. If $(r_1, ----, r_n) \equiv 1$, mod B, the period has n terms; if $(r_1, ----, r_n)$ belongs to exponent K, mod. B, the period has Kn terms.

$$\text{Given } \frac{1}{B} = \frac{p_1}{r_1} + \frac{p_2}{r_1 r_2} + ------ \frac{q_n}{r_1, ---, r_n B}$$

From the method followed in building up this series, $r \equiv q$, mod B

$$r_1 r_2 \equiv q_2, \text{ mod B},$$

and in general, $r_1, ----, r_n \equiv q_n$, mod B.

If $q_n = 1$, $r_1, ----, r_n \equiv 1$, mod B, and there are n terms in the period. In order to obtain the period if q_n is different from one, the series must be continued until $(r_1, ---, r_n)^K \equiv 1$, mod B, and the period will then consist of Kn terms.

The number of terms in the period is independent of the value of the numerator.

$$\text{Let } \frac{A}{B} = \frac{p_1}{r_1} + \frac{p_2}{r_1 r_2} + ---- \frac{q_n}{(r_1, ---, r_n) B}.$$

Then $A r_1 \equiv q_1$, mod B,

and $A r_1 r_2 \equiv q_2$, mod B

and in general $A(r_1,---r_n) \equiv q_n$, mod B.

In order that there may be a period, either q_n , or q_{2n} , etc. must equal A. If $q_n = A$, $A(r_1,----r_n) \equiv A$, mod B, or $(r_1,----r_n) \equiv 1$, mod B. Then, in general, if $q_{Kn} = A$, $A(r_1,---r_n)^K \equiv A$, mod B, and $(r_1,---r_n)^K \equiv 1$, mod B. Thus when q_{Kn} is the first q which equals A, the period would consist of Kn terms, but when this condition exists, $(r_1,---r_n)^K \equiv 1$, mod B, or $\frac{1}{B}$ has a period of Kn terms. Therefore, $\frac{A}{B}$ which is any proper fraction in lowest terms has the same number of terms in its period as $\frac{1}{B}$, or the length of the period is independent of the value of the numerator.

The resolution of $\frac{A}{B}$ into the series $\frac{p_1}{r_1} + \frac{p_2}{r_1 r_2} + \text{———}$ gives a simple method of expressing any fraction as the sum of a number of fractions with unit numerators. $r_1 r_2 --- r_n$ being arbitrary may be chosen in such a way that, when the series is formed, p_1 , $p_2,---, p_n$ will always be unity. If $r_1 r_2,---,r_n$ are given the smallest values possible in order to have p_1 , $p_2 ---- p_n$ unity, the remainders diminish to zero and the series is terminating.

For example:- $\frac{3}{7} = \frac{1}{3} + \frac{1}{3.4} + \frac{1}{3.4.7}$

$$\frac{5}{11} = \frac{1}{3} + \frac{1}{3.3} + \frac{1}{3.3.11} .$$

$\frac{A}{B}$ may also be resolved into a series of fractions whose numerators obey other laws. For instance, r_1 , r_2 ,---,r_n may be so selected that $p_n = p_{n-1} + p_{n-2}$ where p_n , p_{n-1} p_{n-2} , etc. are successive numerators, p_1 being taken equal to 1 and $p_2 = 2$. When the numerators have this form, $r_1 = Bp_1 + 1$, $r_2 = Bp_2 + 1$ and $r_n = Bp_n + 1$ or $r_n = r_{n-1} + r_{n-2} - 1$. The following series show this law of formation.

$$\frac{1}{2} = \frac{1}{3} + \frac{2}{3.5} + \frac{3}{3.5.7} + \frac{5}{3.5.7.11} + \frac{8}{3.5.7.11.17} + ----$$

$$\frac{1}{6} = \frac{1}{7} + \frac{2}{7.13} + \frac{3}{7.13.19} + \frac{5}{7.13.19.31} + ----$$

$$\frac{1}{7} = \frac{1}{8} + \frac{2}{8.15} + \frac{3}{8.15.22} + \quad ----$$

If $r_1 = r_2 = r_3 = ---- r_n$, $\frac{A}{B} = \frac{p_1}{r} + \frac{p_2}{r^2} + \frac{p_3}{r^3} + --- \frac{p_n}{r^n} +$

The series thus obtained is an important special case of the general form since by it, $\frac{A}{B}$ can be expressed to any integral base r, r greater than one. If in this series r has the value 10, the fractions which are known as decimals are obtained. Then since every rational fraction $\frac{m}{n}$, $\frac{m}{n}$ being a proper fraction in lowest terms, may be expressed as a decimal, it is in order to investigate when the decimal will be terminating, periodic, or of unlimited extent.

The necessary and sufficient condition that a decimal terminates is that n must be of the form $2^{\alpha} 5^{\beta}$ where α and β may both be positive integers or either α or β may be zero.

Assume $\frac{m}{n} = \frac{a}{10} + \frac{b}{10^2} + \frac{c}{10^3} + ----- \frac{K}{10^r}$.

The right hand member of the equation may be reduced to an integer by multiplying by the highest power of ten occurring in the denominators, say 10^r. Then $\frac{m}{n}$ must be multiplied by 10^r and the result must also be integral in order to preserve the equality. However, the only way in which $10^r \cdot \frac{m}{n}$ could be integral would be for n to consist of factors of 10 only, or to be of form $2^{\alpha} 5^{\beta}$.

The given condition is also sufficient.

Let $\frac{m}{2^{\alpha} 5^{\beta}} = \frac{a}{10} + \frac{b}{10^2} + \frac{c}{10^3} + -----$ and let r be the highest power of either 2 or 5 occurring in the denominator. Then $\frac{m}{2^{\alpha}5^{\beta}}$ $\cdot 10^{r}$ equals an integer. $10^{r}\left(\frac{a}{10}+\frac{b}{10^2}+\frac{c}{10^3}---\right)$ must also equal an integer, or in other words, the decimal must be terminating.

From the proof given for the previous theorem, it is clear that the highest power of either 2 or 5 occurring in the denominator n, determines the power of 10 used as multiplier and, therefore, determines the length of the period. Several numerical examples will serve to illustrate this statement.

$\frac{3}{160}=\frac{3}{2\ 5}=$ .00625, period five.

$\frac{7}{1000}=\frac{7}{2\ 5}=$ .007 , " three.

$\frac{3}{250}=\frac{3}{2.5}=$ .012 , " three.

If n contains powers of 2 and 5 as well as other factors, the powers of 2 and 5 may be removed, and after a certain number of places, the fraction will have the same mantissa as some fraction with a denominator prime to 10.

$$\text{Let } \frac{m}{n}=\frac{m}{2^{\alpha}5^{\beta}n'}=\frac{a}{10}+\frac{b}{10^2}---\frac{f}{10^{\alpha}}+\frac{s}{10^{\alpha+1}}+\frac{K}{10^{\alpha+2}}+\ ---$$

and assume $\alpha > \beta$. The same proof applies if $\alpha < \beta$ or $\alpha = \beta$.

$$\text{Then } \frac{10^{\alpha}\ m}{2^{\alpha}5^{\beta}n'}= 10^{\alpha}\left(\frac{a}{10}+\frac{b}{10^2}+ ---- \frac{f}{10^{\alpha}}\right)+\frac{s}{10}+\frac{K}{10^2}+ --$$

$$\text{or Integer} + \frac{m'}{n'}=\text{Integer}+\frac{s}{10}+\frac{K}{10^2}+ ----$$

$$\text{Therefore, } \frac{m'}{n'}=\frac{s}{10}+\frac{K}{10^2}+ -----$$

The denominator n' being prime to ten, $\frac{m}{n}$ after a certain number of places will have the mantissa of some fraction whose denominator is free from factors of ten.

For illustration:-

$\frac{13}{45}=\frac{13}{3\ .5}=$ $.2\frac{8}{9}$

$$\frac{1}{28} = \frac{1}{2^2 .7} = .03\frac{4}{7}$$

Having proved that every fraction whose denominator is of the form $2^{\alpha} 5^{\beta}$ is terminating and also that a fraction whose denominator is of the form $2^{\alpha} 5^{\beta} n'$, n' being prime to 10, has the same mantissa as $\frac{m}{n}$ after a certain number of places, the following theorem may be stated:

Any rational fraction $\frac{m}{n}$, n being prime to 10 is periodic. It is assumed that $\frac{m}{n}$ is a proper fraction in lowest terms.

Two proofs of this theorem may be given. If the fraction $\frac{m}{n}$ is considered, there can be only a finite number of different remainders possible when the indicated division is performed. Each place of the remainder can be filled in only a finite number of ways. The product of the number of ways in which each place may be filled will give the number of possible remainders and this must always be a finite number. This rule gives a maximum for the length of the period, but does not always give the length of the period. For instance, the maximum number of places in the period for $\frac{13}{87}$ would be eighty-seven while the actual period has only twenty-eight places.

The second proof for the theorem is as follows:-

Let $\frac{m}{n} = .a\ b\ c\ d\ e\ -----$

$10^{\phi(n)}\frac{m}{n} = 10^{\phi(n)}\ .\,.a\ b\ c\ d\ e\ ----$

But $10^{\phi(n)} \equiv 1, \bmod n.$

Then $10^{\phi(n)}\frac{m}{n} = (Kn + 1)\frac{m}{n} = K + \frac{m}{n}.$

Therefore, a certain period K would be obtained and then that period repeated.

The converse of the theorem just proved, or every periodic decimal is a rational fraction is true except for the case when the period is .999 ----

Let . abcd abcd --- be a periodic decimal. Then $s = \frac{a}{1 - r} =$

$\frac{a \cdot 10^4 + b \cdot 10^3 + c \cdot 10^2 + d\, 10}{10^4 - 1}$ which is a rational fraction. When the period is .999 ----, it is clear that s will equal 1.

The general rules which govern the number of places in the decimal period, considering the fraction as having unity for a numerator will now be considered, and then it will be shown that the length of this period is independent of the value of the numerator.

Given any fraction $\frac{1}{p}$, p being an odd prime, finding the period of this fraction consists simply in finding the exponent to which 10 belongs, modulus p. If 10 is a primitive root of p, the period will be of length p - 1. If 10 is not a primitive root of p, the exponent to which p belongs will be some divisor of ϕ (p) according to Fermat's theorem.

If in the fraction $\frac{1}{n}$, n is the product of different odd primes, the period of $\frac{1}{n}$ is equal to the L. C. M. of the periods of the primes into which n can be resolved. This theorem is proved by the following well-known theorem: If $n = p^{\alpha} q^{\beta} r^{\gamma}$ -- where p, q, r, etc. are different primes and if f, g, h, ---- are the exponents to which a belongs, moduli p^{α} , q^{β} , r^{γ} ---, then t being the L. C. M. of f, g, h ----, $a^t \equiv 1$, mod. n . When a = 10, $10^t \equiv 1$, mod n, or the required period is of length t.

For example, $10^6 \equiv 1$, mod 13, and $10^2 \equiv 1$, mod 11.

Then $10^6 \equiv 1$, mod (11x13) or the period of $\frac{1}{143}$ is six.

If n is of form p^{α} and $\frac{10^s - 1}{p}$ is prime to p, s being exponent to which 10 belongs, modulus p, then the length of the resulting period is equal to the length of the original period multiplied by p .

When 10 is a primitive root of p and also a primitive root of p^2, then 10 belongs to exponent p - 1, mod p, and to exponent $\phi(p^2)$ or p (p-1), mod p^2. From the theory of primitive roots, it is known that in order that 10 may be a primitive root of p and also a primitive root of p^2, then $\frac{10^{p-1}-1}{p}$ must be prime to p.

If 10 is not a primitive root of p, then it must belong to some exponent which is a divisor of p - 1, say f. Then $10^f \equiv 1$, mod p.

I. $10^f = 1 + Kp.$

Raising both sides of the equality to the p th power,

II. $10^{fp} = (1 + Kp)^p = 1 + p \cdot Kp + \text{-----}$

or $10^{fp} \not\equiv 1$, mod p^2, if K is prime to p. But if K is prime to p, from Equation I, $\frac{10^f - 1}{p}$ is prime to p.

If the exponent to which 10 belongs, modulus p^2, is not fp, it must be some divisor of fp. This exponent then must be f, p, some divisor of f as s, or sp.

From Equation I we see that 10 cannot belong to exponent f, modulus p^2, K being prime to p in this equation.

$10^p \not\equiv 1$, mod p, and therefore $10^p \not\equiv 1$, mod p^2. By hypothesis f is the exponent to which 10 belongs, mod p, then $10^s \not\equiv 1$, mod p, and $10^s \not\equiv 1$, mod p^2.

$10^s = X + mp$

Then $10^{sp} = (X + mp)^p = X^p + mp^2 + \text{----}$

$10^{sp} \equiv X^p$, mod p^2.

But $X^p \not\equiv 1$, mod p^2.

Therefore, $10^{sp} \not\equiv 1$, mod p^2.

10^{fp} is then not only congruent to one, modulus p^2, but fp is the exponent to which 10 belongs, modulus p^2, when K is prime to p.

Any primitive root of p^2 is also a primitive root of p^α and the theorem holds for this case. If 10 is not a primitive root of p^2, from Equation II, it is evident that if the theorem holds for p^2, it also holds for p^α.

The length of the period will in any case be independent of the value of the numerator.

If n is any given number, there are always $\phi(n)$ different numbers less than n and prime to n. Then if $\frac{m}{n}$ is any proper fraction in lowest terms, there will always be $\phi(n)$ different fractions with a denominator n having numerators prime to n and less than n. But the $\phi(n)$ different numbers less than n and prime to n form a group in respect to multiplication, modulus n. Therefore, the numerators of the given fractions also form a group in respect to multiplication, modulus n.

For every fraction $\frac{m}{n}$, n prime to 10, 10 will be one of the $\phi(n)$ numbers, if $10 < n$. If $10 > n$, some residue of 10, mod n, will occur. The powers of 10, mod n, form a cyclic sub-group G_1 of the larger multiplication group G.

All the fractions having numerators belonging to G_1 will have a period of the same length as $\frac{1}{n}$.

Assume $10^x \equiv 1$, mod n, or that the period of $\frac{1}{n}$ contains X places. Then $\frac{1}{n} = .a_1 a_2, \text{---} a_x a_1 a_2 \text{---}$. Multiplying by 10 simply changes the decimal point each time one place to the right, or the length of the period for each fraction whose numerator belongs to G_1 is of the same length as that of $\frac{1}{n}$. There is only a cyclical interchange of the numbers composing the period.

An operator of G not in G_1 multiplied by an operator of G_1 will give some distinct element of the group not in the cyclic sub-group. Let K be any operator of G which does not occur in G_1 . Then suppose $\frac{K}{n} = \beta_1 \beta_2 \beta_3 --- \beta_2 \beta_2 ---$. Multiplying by 10 simply moves the decimal point one place to the right, therefore, the multiples of $\frac{K}{n}$ by powers of 10 will have a period of the same length as $\frac{K}{n}$. But it has been assumed that 10 belongs to exponent X mod n, or $10^X \equiv 1$, mod n, therefore $\frac{K}{n} \cdot 10^X$ gives a period of the same length as that of the original period of the cyclic sub-group. If the operators of G are not yet exhausted, another operator not already used may be chosen and the above reasoning repeated. It is then clear, since all fractions with numerators belonging to G_1 have the same period as $\frac{1}{n}$ and all others have the same period as fractions whose numerators belong to G_1 , that the length of the period is independent of the value of the numerator.

A concrete example will serve to illustrate these statements. Assume $n = 39$. Then $\phi(39) = \phi(3) \cdot \phi(13) = 24$ and the fractions having numerators prime to 39 are as follows: $\frac{1}{39}$, $\frac{2}{39}$, $\frac{4}{39}$, $\frac{5}{39}$, $\frac{7}{39}$, $\frac{8}{39}$, $\frac{10}{39}$, $\frac{11}{39}$, $\frac{14}{39}$, $\frac{16}{39}$, $\frac{17}{39}$, $\frac{19}{39}$, $\frac{20}{39}$, $\frac{22}{39}$, $\frac{23}{39}$, $\frac{25}{39}$, $\frac{28}{39}$, $\frac{29}{39}$, $\frac{31}{39}$, $\frac{32}{39}$, $\frac{34}{39}$, $\frac{35}{39}$, $\frac{37}{39}$, and $\frac{38}{39}$.

In this case, the cyclic sub-group G_1 of the numerators will be 10, 22, 25, 16, 4, 1.

$\frac{10}{39} = .256416$ --- $\frac{16}{39} = .410256$ ---

$\frac{22}{39} = .564102$ --- $\frac{4}{39} = .102564$

$\frac{25}{39} = 641025$ ---- $\frac{1}{39} = .025641$ ---

Taking 2 as an operator, another set of six numbers would be given by the fractions $\frac{20}{39}$, $\frac{5}{39}$, $\frac{11}{39}$, $\frac{32}{39}$, $\frac{8}{39}$, $\frac{2}{39}$. The periods would be a cyclical interchange of the numbers .512820.

Using 7 as an operator, the fractions would be $\frac{31}{39}$, $\frac{37}{39}$, $\frac{19}{39}$, $\frac{34}{39}$, $\frac{28}{39}$, $\frac{7}{39}$, the periods of which would be formed by the cycle .794871.

The remaining set of fractions is $\frac{23}{39}$, $\frac{35}{39}$, $\frac{38}{39}$, $\frac{29}{39}$, $\frac{17}{39}$, and $\frac{14}{39}$ with the cyclic period .589743.

The number of different cyclic sets, or periods of the same length for each denominator n is equal to $\phi(n)$ divided by the exponent to which 10 belongs, modulus n. This is true since $\phi(n)$ gives the number of fractions having numerators in the multiplication group G and the exponent to which 10 belongs gives the number of these fractions in each cyclic set.

The number 10 belongs to exponent 6, modulus 39, and $\phi(39)$ is 24. Therefore, there are four different cyclic sets for n = 39, as was seen in the previous example. When n = 7, 10 belongs to exponent 6, mod n, and there is only one cyclic set. In this case the periods of $\frac{1}{7}$, $\frac{2}{7}$, $\frac{3}{7}$, $\frac{4}{7}$, $\frac{5}{7}$, $\frac{6}{7}$ consist of the numbers .142857 arranged in different order. From this we see that 142857 multiplied by any number less than seven will always give the same numbers, only in a different order.

The operation of finding the numbers composing the different cyclic sets of the $\phi(n)$ fractions is very much shortened by the fact that, in general, the different cyclic sets, or periods, occur as complements of each other.

Let $\frac{X_0}{n}$ and $-\frac{X_0}{n}$ be any two fractions differing only in sign.

If $\frac{X_0}{n}$ = $.a_1 a_2 a_3$ ---- and $-\frac{X_0}{n}$ = $.b_1 b_2 b_3$ ----, then the sum of the two fractions must equal zero and, therefore, a_1 is the complement of b_1 , a_2 the complement of b_2, etc.

Since -1 and +1 always occur among the $\phi(n)$ numbers of G, the $\phi(n)$ fractions whose numerators differ only in respect to sign, mod n, have periods occurring in complementary pairs. Then if a period has been obtained, its complementary period is found by subtracting each digit of the first period from nine. For instance, in the example already given for n = 39, the period of the fractions having numerators 10, 22, 25, 16, 4, 1 is the cyclic set .256416. Then the period for the complementary fractions $\frac{29}{39}$, $\frac{17}{39}$, $\frac{14}{39}$, $\frac{23}{39}$, $\frac{35}{39}$, and $\frac{38}{39}$ will be the cyclic set obtained by subtracting the numbers 2, 5, 6, 4, 1, 0 from 9 or the cyclic set .743589.

If 10 is a primitive root of n, then -1 is in the cyclic subgroup G_1 . The period is in this case of even length and the second half of the period is obtained by subtracting the digits of the first half from nine. The period is of even length since G_1 then contains an operator of order two and the order of the operator must divide the order of G_1 . If -1 is not in G_1 , then the index of G_1 under G must be even since the periods are complementary.

Chapter III

The complex number systems which we shall consider in Chapter III are quadratic complex systems restricted to certain integral domains. The term integral domain will be understood, as usual, to mean a totality of integral elements which is invariant with respect to addition, subtraction, and multiplication. The special domains to which the quadratic complex systems in this case will be restricted are algebraic domains formed by adjoining an irrational algebraic number to the ordinary integral domain.

An algebraic integer is a root of the equation $x^n + a_1x^{n-1} + a_2x^{n-2} + ---- a_n = 0$ where a_1, a_2, ---, a_n are rational integers. An algebraic quadratic integer is a root of the above equation when $n = 2$. All algebraic quadratic integers are of form $x + y\sqrt{m}$ when $m \equiv 2$ or 3, mod. 4, and of form $x + y\frac{1 + \sqrt{m}}{2}$ when $m \equiv 1$, mod. 4. It will be assumed that m is not divisible by any square greater than 1 and that x and y are integral.

An integral algebraic domain can then be formed by adjoining $\sqrt{-6}$ to the ordinary integral domain. This algebraic integral domain will be denoted for brevity by w and some of the properties of a quadratic complex system in it will first be discussed.

It is well known that numbers which are primes in one domain may be composites in another and also that numbers can be resolved into prime factors in more than one way in some domains. The quadratic complex system in w will be considered with special reference to these two points.

All ordinary primes of the form 24Z + 1 and 24Z + 7 are composite in w.

From the theory for binary quadratic forms, it is know that D, the determinant of the form, must be a quadratic residue of m, where m is the number to be represented by the form. The number -6 is a quadratic residue of primes of form 24Z + 1, 24Z + 7, 24Z + 5, and 24Z + 11. There are only two reduced forms for D = -6, namely, $x^2 + 6y^2$ and $2x^2 + 3y^2$. The form $2x^2 + 3y^2$ will not give complex factors in w so need not be considered. Then all ordinary primes which are composite in w must be represented by $x^2 + 6y^2$.

If m is the prime to be represented, $m = x^2 + 6y^2$ and, therefore, $x^2 \equiv m$, mod. 6. But $x^2 \equiv 1$, mod. 6, and, therefore, all ordinary primes represented by $x^2 + 6y^2$ are of form 24Z + 1 and 24Z + 7 and only primes of these forms can be factored.

$(1 - \sqrt{-6})(1 + \sqrt{-6}) = 7$; $(5 - \sqrt{-6})(5 + \sqrt{-6}) = 31$

$(7 + 2\sqrt{-6})(7 - 2\sqrt{-6}) = 73$; $(5 - 3\sqrt{-6})(5 + 3\sqrt{-6}) = 79$

Primes of the forms 24Z + 1 and 24Z + 7 can be resolved into their complex factors in only one way.

There are four representations of $x^2 + 6y^2$ which give the same prime m. There are two solutions of the congruence, $n^2 \equiv -6$, mod. m, and two substitutions transforming $x^2 + 6y^2$ into an equivalent form. Therefore, a prime number can be represented by $x^2 + 6y^2$ in only one way and consequently primes of form 24Z + 1 and 24Z + 7 can be resolved into prime complex factors in only one way.

In order that a composite rational integer may be resolved into complex factors only in w, it must be of the form, $m = a^{(\alpha)} B^{(\beta)} c^{\gamma} d^{\delta}$ where a represents primes of the form 24Z + 1 and 24Z + 7 and (α) the number of primes in a; B primes of the form 24Z + 5 and 24Z + 11 and (β) the number of primes in B; c = 2, d = 3, $(\beta) + \gamma + 8$ is an even number and $\gamma + 8 \ngtr (\beta)$.

The determinant of the form or - 6 must be a quadratic residue of any composite number m properly represented by $x^2 + 6y^2$ and, therefore, must be a quadratic residue of every prime factor of m. $x^2 + 6y^2 = m$, or m must be a quadratic residue of 6. The number -6 is a quadratic residue of primes of form 24Z + 1, 24Z + 7, 24Z + 5, and 24Z + 11. Primes of form 24Z + 1 and 24Z + 7 are quadratic residues of 6 while primes of the form 24Z + 5 and 24Z + 11 are non-quadratic residues of 6. Then any number $m = a^{(\alpha)} B^{(\beta)}$ is properly representable by $x^2 + 6y^2$, or may be resolved in complex factors only in w, (β) being even, since an even number of non-quadratic residues is a quadratic residue.

I $2x^2 + 3y^2 = (x\sqrt{2} + y\sqrt{-3})(x\sqrt{2} - y\sqrt{-3})$

II Then $2(2x^2 + 3y^2) = \sqrt{2}(x\sqrt{2} + y\sqrt{-3}) \cdot \sqrt{2}(x\sqrt{2} - y\sqrt{-3})$

$= (2x + y\sqrt{-6})(2x - y\sqrt{-6}) = (Z + y\sqrt{-6})(Z - y\sqrt{-6})$, Z being an even number.

III Also $3(2x^2 + 3y^2) = \sqrt{3}(y\sqrt{3} + x\sqrt{-2}) \cdot \sqrt{3}(y\sqrt{3} - x\sqrt{-2})$

$= (3y + x\sqrt{-6})(3y - x\sqrt{-6}) = (Z + x\sqrt{-6})(Z - x\sqrt{-6})$

Primes 2 and 3 are not represented by $x^2 + 6y^2$ and are not properly represented by $2x^2 + 3y^2$. Equations II and III then show that neither 2^{α}, 3^{β}, $2^{\gamma} 3^{\delta}$ are properly represented, or have complex factors in domain w. Primes of the form 24Z + 5 and 24Z + 11, however, are properly represented by $2x^2 + 3y^2$ and an inspection of Equations II and III will show that the product of any such prime or B, by either 2 or 3 will give a number having complex factors in w. Since neither 2^{α}, 3^{β}, or $2^{\gamma} 3^{\delta}$ have complex factors in w, but 2 B and 3 B have such factors, $\gamma + \delta$ cannot be greater than (β). $(\beta) - (\gamma + \delta)$ must be even since all primes of form 24Z + 5 and 24Z + 11 are non-quadratic residues, modulus 6, $2(\gamma + \delta)$ is even. Then $(\beta) + \gamma + \delta$ is an even number.

$25330 = 7 . 5. 11. 29. 2 = (1 - \sqrt{-6})(1 + \sqrt{-6})(7 - \sqrt{-6})(7 + \sqrt{-6})$
$(2 + 3\sqrt{-6}) x (2 - 3\sqrt{-6})$.

Any composite rational integer $m = a^{(\alpha)} B^{(\beta)} c^{\gamma} d^{\delta}$ representable by the form $x^2 + 6y^2$ can be resolved into its prime complex factors in more than one way providing it contains at least two different prime factors of the form 24Z + 5 and 24Z + 11 so that $(\beta) - (\gamma + \delta) \nless 2$.

Let Kx represent the composite rational integer, x representing product of prime factors of form 24Z + 5 and 24Z + 11, K the product of all other prime factors. K can be resolved into complex prime factors in only one way because primes of form 24Z + 1 and 24Z + 7 are resolvable into complex factors in only one way and 2^{α}, 3^{β} or $2^{\gamma} 3^{\delta}$ have no complex factors in the domain. It has also been shown in the previous theorems that no prime of the form 24Z + 1 and 24Z + 7 combined with either 2 or 3 or with a single prime of the form 24Z + 5 or 24Z + 11 can be resolved into complex factors only in w. 2 B and 3 B, where B is any prime of the form 24Z + 5 or 24Z + 11, can be resolved into complex factors in only one way since 2 and 3 have no complex factors in w. From these statements it is evident that the factoring in different distinct ways must then depend only on the factors of x.

A number x can be represented or resolved into its complex factors in w in 2^{u-1} ways where u represents the number of different prime factors. These u - 1 representations will be distinct since primes of form 24Z + 5 and 24Z + 11 are primes in w.

$55 = (1 + 3\sqrt{-6})(1 - 3\sqrt{-6})$

$55 = (7 + \sqrt{-6})(7 - \sqrt{-6})$

$145 = 29.5 = (11 + 2\sqrt{-6})(11 - 2\sqrt{-6}) = (7 + 4\sqrt{-6})(7-4\sqrt{-6})$

$385 = 7 \cdot 5 \cdot 11 = (13 + 6\sqrt{-6})(13 - 6\sqrt{-6}) = (1 + 8\sqrt{-6})$
$(1 - 8\sqrt{-6}) = (19 - 2\sqrt{-6})(19 + 2\sqrt{-6}) = (17 + 4\sqrt{-6})$
$(17 - 4\sqrt{-6})$. These four representations or four sets of complex factors for 385 are not distinct. $(13 + 6\sqrt{-6})(13 - 6\sqrt{-6})$ and $(1 + 8\sqrt{-6})(1 - 8\sqrt{-6})$ both have the same prime complex factors, namely, $(1 + \sqrt{-6})(1 - \sqrt{-6})(7 + \sqrt{-6})(7 - \sqrt{-6})$. Likewise $(19 - 2\sqrt{-6})(19 + 2\sqrt{-6})$ and $(17 + 4\sqrt{-6})(17 - 4\sqrt{-6})$ have the same prime complex factors $(1 + \sqrt{-6})(1 - \sqrt{-6})(1 + 3\sqrt{-6})(1 - 3\sqrt{-6})$.

The primes in the quadratic complex system in w comprise:-

(1) The real number primes which are congruent to 5, 11, 13, 17, 19, and 23, modulus 24, and the primes 2 and 3.

(2) The number $\sqrt{-6}$ which corresponds to $x = 0$ and $y = 1$ in $x + y\sqrt{-6}$.

(3) The numbers $x + y\sqrt{-6}$ of which the norms are the ordinary primes congruent to 1 and 7, modulus 24. Also the numbers $x + y\sqrt{-6}$ whose norm contains only two prime factors, each such prime not being representable by $x^2 + 6y^2$.

Bibliography

Cahen, E.

Eléments de la Théorie des Nombres. 1900.

Cauchy, M. Augustin

Operations exécutées avec deux especes de chiffres, les un positifs, les autres négatifs.

Comptes Rendus - July - Dec. 1840.

Chrystal, G.

Algebra, Vol. I.

MacMillan and Co., 1889.

Conant, Levi Leonard

The Number Concept.

MacMillan and Co., 1896.

Cooper, George

Octimal Notation.

The Whitaker and Ray Company. 1902.

Eissfeldt, Ferdinand

Invention of a New Numerical System.

Analyst, Vol. II, 1875.

Fontene, M. G.

Sur les Entiers Algébriques de la Forme $X + Y\sqrt{-5}$.

Nouvelles Annales de Mathématiques, Vol. III, 1903.

Johnson, W. Woolsey

Octonary Numeration.

Bulletin of the New York Mathematical Society, Vol. I, 1891 - 1892.

Nystrom, John W.

Elements of Mechanics containing an Appendix on Duodenal Arithmetic.

Porter and Coates, Philadelphia. 1875.

Taylor, Alfred

Octonary Numeration and its Application to a System of Weights and Measures.

Proceedings of the American Philosophical Society; Vol. XXIV, 1887.

Western, A. E.

Certain Systems of Quadratic Complex Numbers.

Transactions of Cambridge Philosophical Society, Vol. XVII, 1899.

Spotted something that isn't quite right?

Let us know and we'll fix it for future reprints.
And as a thank you for your diligence, we'll
send you a free Sothis book of your choice.

publisher@sothis.press

This page intentionally left blank

Made in United States
Cleveland, OH
06 November 2026

47793875R00023